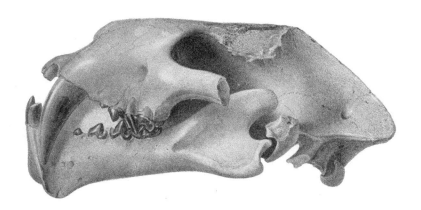

EXPEDITION
TO THE JOHN DAY RIVER
⊫ IN 1878 ⊰⊰

an excerpt from **THE LIFE OF A FOSSIL HUNTER** by Charles H. Sternberg

Front cover:
Charles Sternberg and his three sons prepare for a day of fossil hunting
wearing head gear to ward off insects.

Background photo:
Fossil bearing cliffs. Middle John Day exposure.

Book Design: Mark MacKay
© 2001, 2016 Discover Your Northwest

Published by Discover Your Northwest
164 S Jackson Street
Seattle, WA 98104   (877) 874-6775
discovernw.org

National Park Service
John Day Fossil Beds National Monument
www.nps.gov/joda/

Special thanks to Indiana University Press

THE FRONTIER BENEATH THE SURFACE

AN INTRODUCTION TO STERNBERG'S *Expedition to the John Day River in 1878*

Charles H. Sternberg (1850-1943) was a young man of 28 when he came to look for fossils near the John Day River, early in his 70-year career as an internationally acclaimed fossil collector.

When he came to Oregon in 1878, he was working for Edward Drinker Cope, one of the most prolific and important paleontologists of all time. In fact, Sternberg had been on another Oregon mission during the summer of 1877, when Cope sent him to search for fossils in an Ice Age lake bed southwest of the John Day region. Sternberg named the site Fossil Lake. The llama, horse, dog, beaver and other animals he found there were probable descendents of older species he found among the colorful John Day claystones the following summer.

Sternberg had first experienced the excitement of finding fossils not far from home as a boy in upstate New York. Near Cooperstown, he explored limestone formations from which he cut shells to give his mother. Later, he and his brothers would discuss fossils discovered several miles away where the Erie Canal was being widened. A highlight of his teenage years in Kansas was discovering a pile of rounded concretions which when split revealed nearly perfect impressions of deciduous leaves, notably sassafras.

Sternberg went on to establish a highly successful fossil collecting enterprise with his three sons, George, Charles and Levi. Together, they left a legacy that encompassed numerous landmarks in Kansas, Wyoming, Texas, Montana, Southern Alberta, and even Patagonia. Through decades of field experience, Sternberg and his sons established many techniques still practiced today such as plaster jacketing of specimens for protection during transport. Their well regarded work served the scientific community and made a highly valuable contribution to American paleontology.

BLUE MOUNTAIN REGION

OF

NORTHEASTERN OREGON

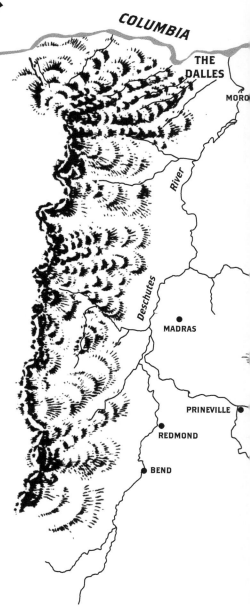

COLUMBIA

THE DALLES

MORO

River

Deschutes

MADRAS

PRINEVILLE

REDMOND

BEND

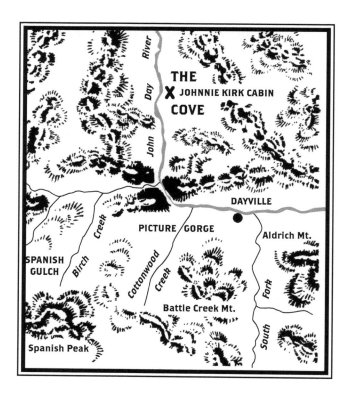

John Day River

THE COVE

X JOHNNIE KIRK CABIN

DAYVILLE

PICTURE GORGE

SPANISH GULCH

Birch Creek

Cottonwood Creek

Aldrich Mt.

South Fork

Battle Creek Mt.

Spanish Peak

The Cove Area Enlarged

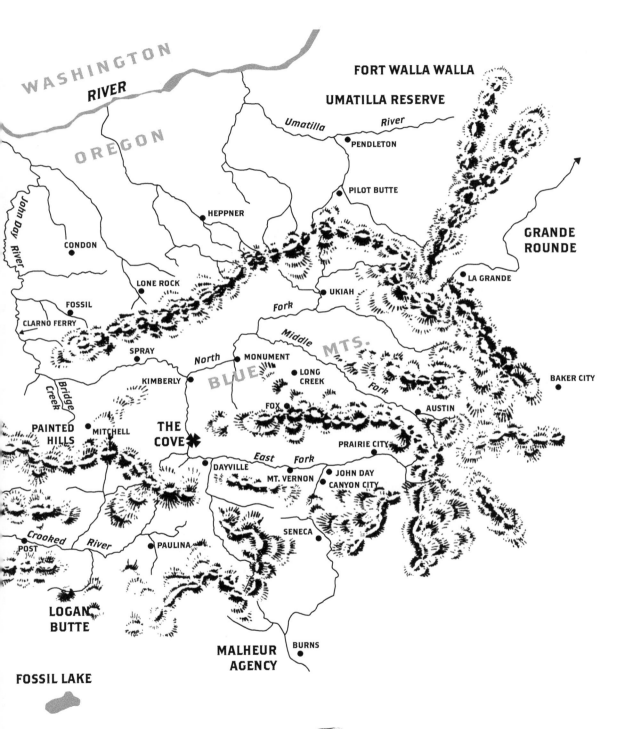

WASHINGTON

OREGON

RIVER

John Day River

FORT WALLA WALLA

UMATILLA RESERVE

Umatilla   River

PENDLETON

PILOT BUTTE

HEPPNER

CONDON

LONE ROCK

FOSSIL

CLARNO FERRY

SPRAY

Bridge Creek

KIMBERLY

North

Fork

MONUMENT

Middle

BLUE   MTS.

Fork

LONG CREEK

FOX

UKIAH

GRANDE ROUNDE

LA GRANDE

BAKER CITY

AUSTIN

PAINTED HILLS

MITCHELL

THE COVE

DAYVILLE

East   Fork

MT. VERNON

JOHN DAY

CANYON CITY

PRAIRIE CITY

Crooked   River

POST

PAULINA

SENECA

LOGAN BUTTE

MALHEUR AGENCY

BURNS

FOSSIL LAKE

Expedition to the John Day River in 1878

INTRODUCTION   III

The vivid account provided here of young Sternberg's visit to places now within the Sheep Rock Unit of John Day Fossil Beds National Monument, is but one chapter is his original memoirs, "The Life of a Fossil Hunter", first published in 1909. His expedition to the John Day River took him places that people continue to visit today-the intriguing Blue Basin which Sternberg calls "the Cove," and Picture Gorge, then called by him "the Grand Coulee." The chapter captures a sense of remoteness and adventure that visitors can still experience.

Sternberg's 1878 expedition coincides with an outbreak of conflict between several Northwest Indian tribes, the U. S. military, and eastern Oregon settlers. In his account, Sternberg uses language that would be considered unacceptable today. One journalist has suggested that perhaps Sternberg exaggerated the hostilities he witnessed to discourage rival fossil collectors from venturing to the area. Regardless of his motivation, Sternberg's account is a genuine artifact of history, and the original text has been maintained in this reprinting.

Sternberg's somewhat accidental account of the Bannock War reflects the tensions of the frontier, although by 1878, Oregon had been a state for nearly 20 years. Sternberg's story overall, however, takes the reader much farther back in history, beyond what is measurable in human time scales. His expedition explores the frontier beneath the surface of human events and its geography.

Though Sternberg understood the fossils he found on his John Day expedition represented life after the dinosaur extinctions, he was not aware that he was working in one of the most complete records for the "Age of Mammals." Sternberg was part of a generation that sought primarily large, charismatic fossils. Skulls were most highly prized, engendering leagues of "head hunters" who often overlooked other material. Sternberg was farsighted for his time, recognizing the importance of "postcrania," although he still probably missed many smaller fossils. By contrast today, all evidences of past life, from microorganisms to gigantic skeletons are collected in an effort to reconstruct entire ecosystems.

Among the important specimens Sternberg collected in the John Day region were the skull of *Pogonodon platycopis*, a large cat-like carnivore from a now extinct lineage, and an early camel which Cope named in his honor, *Paratylopus sternbergi*. These and many other fossils collected by Sternberg for Cope are housed in the American Museum of Natural History in New York.

Much of Sternberg's work for Cope was part of the early government sponsored surveys of the west, overseen by Ferdinand Hayden, director of the U. S. Geological and Geographical Survey of the Territories which later became the U. S. Geological Survey. The "Hayden Surveys" identified significant fossil bearing formations on public lands, many of which today are still held in the public trust.

Since John Day Fossil Beds National Monument was established in 1975, a program of routine fossil prospecting has been implemented to protect fossils as they weather out of the rock and to collect other associated data. These are added to the monument's museum collection and preserved for the purposes of scientific research and education.

Between Sternberg's 1878 expedition and the establishment of the monument, several other scientific expeditions took place, which disseminated fossils from the John Day region to numerous institutions across the country. These include the Smithsonian Institution, the Yale Peabody Museum, the University of California Museum of Paleontology at Berkeley, the University of Washington Burke Museum, and the museum at Princeton University.

A long history of surveys and exploration, driven by dynamic personalities, have mapped and named most of the features on America's surface, both living and non-living, but it remains the work of paleontology to continue to explore the frontier, both beneath the surface and through time. Emerging from John Day Fossil Beds is a story of tropics changing to sagebrush steppe, animals crossing continents, mountain building, river carving, and volcanic outpourings.

Charles H. Sternberg was one of many who will play a part in revealing the unfolding story of landscapes and species now long gone.

Jennifer Chapman, Park Ranger
John Day Fossil Beds National Monument

CHARLES H. STERNBERG 1850 -1943

During the winter of 1877-'78 I camped on Pine Creek, Washington, exploring the swamps in the neighborhood and fighting against water to secure specimens. We had dug a large shaft down to the bed of gravel, twelve feet below the surface, in which bones were to be found, but every morning we found that the hole had filled with mud and water over night, and we had to spend hours bailing it out. When we finally got it clear again, we had little time or strength left for securing fossils. This performance had to be repeated day after day, and of course the farther we excavated, the more water there was to be bailed out. I don't think that we were dry a single day that winter. But luckily the water was warm, and we did not suffer from colds.

On the twenty-third of April I started with a team and wagon from Fort Walla Walla, accompanied by my two assistants, Joe Huff and "Jake" Wortman, the latter at that time an intelligent young man from Oregon, who had been introduced to me the winter before by my brother, Surgeon George M. Sternberg, at that time post surgeon of Fort Walla Walla. During the past six months Wortman had been my guest at my camp on Pine Creek. Afterwards he became known to science as Dr. J. L. Wortman.

We skirted the Blue Mountains in a southwesterly direction, traveling through the beautiful wheatfields of that fertile region; and striking south at Cayuse Station on the Umatilla Reserve, we climbed the long slopes of the mountains and plunged down into the Grande Rounde, once the bed of an ancient lake, but now a lovely valley nestling among the hills. From this point we drove south to Baker City, and leaving behind us the jagged peaks of the Powder River Mountains, struck the John Day River at Canyon City.

On the second of May we camped on the other side of the mountains in a large meadow. The boys went hunting and got a deer. On the third, our road led us again through rugged mountains, covered in places with ice, and we had to cut footholds for our horses, as they were smooth-shod. We passed through a large mining gulch, where men were at work placer-digging for gold. The whole surface of the country had been dug over, and was disfigured with holes and ditches and heaps of earth.

On the fifth of May, after passing through Canyon City, we started for the John Day Basin. It snowed nearly all day. On the road we met a man who told us of a rich fossil leaf locality, on the Van Horn ranch; and after a sixteen-mile drive we found the place and secured some very fine specimens. The leaf impressions were found in a soft, shaly clay-stone, and were very abundant, representing well-preserved Tertiary flora. That night we feasted on a large salmon trout which I caught in an irrigation ditch.

On the sixth (I am following my notebook) we worked all day. I collected two hundred specimens, and Mr. Wortman eighty-five. They were all very fine, and represented the oak, the maple, and other species. I secured some fish vertebrae also. This is another case in which I lost credit for early discoveries. I was told by Professor Cope, a few years before his death, that these specimens had never been examined.

In this same locality there is a bed of rock so light that it floats. I threw a large mass of it at some object in the water, and was amazed to see it float off down the stream. It was the first time that I had ever seen a rock lighter than water.

On the seventh of May, after a journey of fifteen days from Walla Walla, we reached Dayville, a mile below the crossing of the South Fork of the John Day River. One of the first men I met was a certain Bill Day, whom I soon after hired as assistant. He had for years been making collections of the fossil vertebrates here, usually sending them to Professor Marsh. I was able to secure a large and fine collection from him and another mountain man, a Mr. Warfield, who had also spent much time collecting fossils. Both men had been employed by Professor Marsh during his expedition in this region, and were very careful workmen.

We camped on Cottonwood Creek and prepared to pack into the Basin, or Cove as it has been called. For a hundred and fifty miles of its course, the John Day flows east, skirting the Blue Mountains, but here at Cottonwood or Dayville, it has turned north and cut a great canyon, four thousand feet deep, through the heart of the mountains, the so-called Grande Coulée, since known as the Picture Gorge. At the foot of this canyon, the mountains swing away from the river in a great horseshoe bend, closing in upon it again several miles below. This amphitheater, three miles wide and thirteen long, is a scene of surprising beauty. The brilliantly colored clays and volcanic ash-beds of the Miocene of the John Day horizon paint the landscape with green and yellow and orange and other glowing shades, while in the background, towering upward for two thousand feet, rise rows upon rows of mighty basaltic columns, eight-sided prisms, each row standing a little back of the one just below, and the last crowned with evergreen forests of pine and fir and spruce. But no pen can picture the glorious panorama.

Ever since Cretaceous times, when a quiet inland sea laid down the thousand feet of Kansas chalk, here in the John Day region vulcanism has held sway; almost until to-day. Indeed I have often seen the summit of old Mount Hood wreathed with menacing clouds of smoke, as if she were preparing to pour forth again her floods of molten lava and devastate the region.

When volcanic action first began, great masses of ashes must have been thrown out over the country, settling in the lakes and covering the remains of animals which had been accumulating there for ages. Then floods of lava, one after another, poured out over the forests, until they lay buried beneath two thousand feet of volcanic rock. Where did this immense mass of molten rock come from, and how? A dike crosses the Basin, and for fifteen miles the basaltic columns lie along its edges like cordwood; so we know that some of the lava at least was squeezed up out of the earth's crust through narrow cracks.

I remember once, as I was standing with Uncle Johnnie Kirk, the hermit of the Cove, in front of his cabin, he pointed to the basaltic cliffs that towered above us, and observed gravely, "All vegetable matter." He had found at the base remains of the forests which the lava had engulfed, and had concluded that the whole mass represented similar remains.

Before moving the outfit into the fossil beds I took my pony and started off to spy out the land. Following a horse trail that led up the gentle slope west of the canyon represented in Dr. Merriam's picture of the Mascall Beds I reached a tableland, which proved to be the divide between Cottonwood and Birch creeks. Here I found that the trail leading down to the mouth of Birch Creek was very steep - one could have greased one's boots and slid the whole distance of several hundred feet. I was afraid to ride down and led my pony, but I soon learned that an Oregon pony has long, well-developed legs and can climb up and down better than I could myself.

When I reached the river at the mouth of the Grande Coulée, I found to my dismay that all the rich-looking green and brown fossil beds were on the other side, where the amphitheater which I have mentioned is cut out of the flank of the mountains. As a boy I had learned to swim dog-fashion, and as the river was not over thirty or forty feet wide, and I was determined, after coming so far, to find some fossils and a good camping ground, I decided to strip, jump out as far as I could, and paddle the rest of the way across.

No sooner thought than done. In I sprang, discovering too late that I had reckoned without my host and that the river, which had been penned in for miles by the walls of the canyon, was here flowing away from its prison with amazing swiftness and power. My weak little body was as helpless as a straw in its grasp: down I went, and striking a boulder at the bottom, was flung up five feet into the air, I took in breath and closed my mouth as I went down again; tossing me hither and thither like a cork, beating me against rocks and hurling me high into the air, the river bore me swiftly on, until at last, thank God! it tired of its toy, and threw me to one side into deep water, under a willow whose welcoming branches I eagerly clasped. There I hung until I had regained my strength enough to pull myself out.

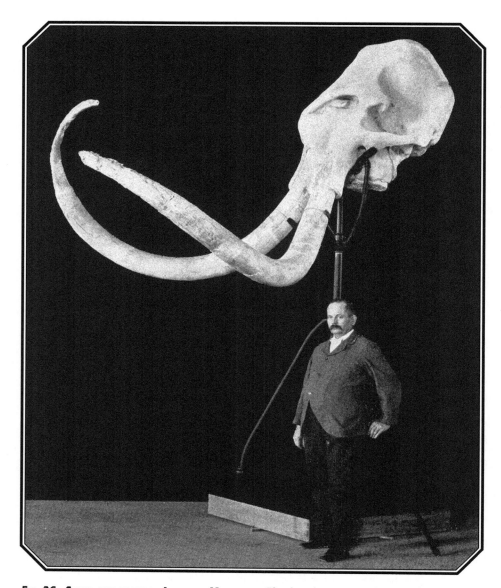

FIG. 26.-SKULL AND TUSKS OF IMPERIAL MAMMOTH, *Elephas imperator*. In American Museum of Natural History. In this portrait Sternberg is pictured with an "Ice Age" mammoth, a fairly recent resident of North America. At John Day Fossil Beds, Sternberg may have found fossils of older elephant-like animals which became extinct before mammoths migrated to this continent.

But the fossil vertebrates of the John Day beds were still across the river and the questions which I had crossed the mountains and risked my life to answer were still waiting for replies. Unwilling to return home beaten I walked up and down the river shore, and was delighted to find an old boat caught in a pile of driftwood. I dug it out with my bare hands, only to find that its seams had parted and that its bottom was as full of holes as a sieve. Not dismayed, I found a bed of sticky clay with which I calked my ship, and venturing again into the flood, managed to get to the other shore before the boat sank.

I found a place to camp lower down, at the mouth of a canyon which opened out into the level country, and on a little creek that ran in front of Uncle Johnnie's cabin. I was very well pleased with my explorations in the fossil beds also, for I found the skull of an Oreodon, a hog-like creature which, judging from the abundance of skulls and skeletons, must have lived in droves during the time when this rock was being deposited in the lakes of this region. These animals were herbivorous in habit. Uncle Johnnie always referred to them as bears. He often brought a skull into camp with the remark, "Here's another bar's head. I've killed hundreds of 'em in ole Virginia."

I returned to camp much elated, and was planning to pack the outfit into the Basin the next day, when to my disgust Joe Huff, who owned the horses, refused to pack them, as he did not want to run the risk of injuring them. It was useless to tell him that he had been hired to do what I wanted, etc.; he was not to be moved. So I paid him off, and saw him start for his home near Moscow, Idaho, riding bareback. I felt sorry for him, but he had a stubborn fit on, and there was no doing anything with him. After I had hired Bill Day, he wanted me to overlook the past and re-employ him, but it was too late then.

I suppose Bill Day must have weighed about one hundred and eighty pounds, but he was an expert hunter and a keen observer. He owned a herd of ponies and furnished me with all that I wanted, and as he knew every inch of the fossil beds and all the best camping grounds, his services were invaluable. He kept our larder supplied with venison, also. I think my success in that region was largely due to his assistance. I was also indebted to a Mr. Mascall, a man who lived on the second bottom of the river. He had an extra log cabin behind the one he lived in, and he let us use it as a storeroom for our extra supplies of food and for our fossils, when we began to secure them.

This Mr. Mascall had a wife and daughter, and when we came in from the fossil beds, after several weeks of camping out, it seemed almost like coming home to be able to put our feet under a table, eat off stone dishes, and drink our coffee out of a china cup, and to sleep on a feather bed instead of a hard mattress and roll of blankets. Then Mr. Mascall was a good gardener, and always had fresh vegetables, a most enjoyable change from hot bread, bacon, and coffee. I shall not soon forget his hospitality.

When all was ready, we were taken across the river in Mr. Mascall's boat, swimming our horses. Then the packs were adjusted, and the wearisome climb up the face of the mountains began. It usually took us half a day to reach the summit. Then we climbed down steep slopes and over spurs of the hills, until we reached Uncle Johnnie Kirk's hospitable cabin, a 12 x 14 structure of rough logs with a shake roof. He kept bachelor's hall and lived all alone, except when some cowman or fossil hunter came along. We pitched our tent near his house.

Not far away there was a tract of bad lands, called the Cove, the largest in the John Day Basin, covering, I should judge, a section of land. It was cut into the usual fantastic forms, peaks, ridges, and battlements, and slender spires sometimes a hundred feet high, and as thickly clustered as those of some old Gothic cathedral. Their summits were crowned with hard concretions, which protected their almost perpendicular sides from destruction by the elements.

The drainage canals spread out through this territory like the ribs of a fan, converging at the entrance, and woe to the man who chanced to be caught in one of them during a rain, for the steep slopes shot the water down into them with such amazing rapidity that before he could turn around he would be engulfed in fathoms of water. We always climbed up to some high point the minute we heard the rain strike the rocks above us, and waited until the storm was over and the water had run out. A ditch containing twenty feet, sometimes, of water would dry up as soon as it stopped raining, so steep was the slope of its bed.

I was continually impressed in this region by the power of running water. Not only is this manifested in the mighty canyons which have been carved out during the course of ages from the solid rock, but I stood transfixed with astonishment once, at the mouth of the little creek in front of Uncle Johnnie's cabin, on finding it dammed by a mass of basaltic rock, weighing at least twenty tons, which had been brought from its native hills, three miles away, by a flood of water, and left stranded here. All the side canyons that empty into the John Day River have dumped their loads of boulders there, in some places damming the stream or creating a series of rapids.

I soon found that all the ground in the fossil beds which was easy to get at had been gone over. Here and there we would run across a pile of broken bones and a hole from which a skull had been taken. When I asked Bill what he had meant by leaving the bones of the skeleton behind, he answered, "We were only looking for heads, though we sometimes saved knucks and jints." This accounts for the scarcity of skeletons among the first collections made. I saw to it that my party

Fig. 27.–**FOSSIL-BEARING CLIFFS.** (After Merriam.) Upper John Day exposure.

Fig. 28.–**FOSSIL-BEARING CLIFFS.** (After Merriam.) Middle John Day exposure.

should care for every bone discovered.

I realized then, that if we were to make our expedition a success, we should have to climb where no one before us had dared to go. It was a serious matter to scale those almost perpendicular heights; one took one's life in one's hand in attempting it. They were, of course, entirely bare of vegetation, and where the slope was not too steep, they were covered with angular fragments of rock which rolled from under one's feet and were likely to send one flying into the gorge below. But I laid the situation before my two men, explaining to them that unless they were willing to face the danger, we should have to give up the expedition, as we had explored the safe ground without results; and they courageously agreed to follow where I led.

So every morning we started out for a day of perilous enterprise, each with a collecting bag over his shoulder and a well-made pick in hand. The latter was used not only for digging out fossils, but was absolutely indispensable as an aid in climbing, and as an anchor in case we began to slip. We were never sure when we left camp in the morning that we should all meet there at night, since a single misstep on those cliffs would mean death or worse than death on the pitiless rocks below; but every day we gained confidence and grew more skillful in the use of our picks.

Far above the pick-marks of the fossil hunters who had preceded us, far above the signs of the mountain sheep that inhabited these wilds, we made our way, cutting niches for our feet as high above us as we could reach, and drawing ourselves up with bodies pressed to the rock. At each niche we rested, and scanned the face of the cliff for the point of a tooth or the end of a bone, or for one of those concretions, among the thousands that everywhere topped the pinnacles or projected from the rocky slopes, whose skull-shaped form revealed the treasure that was hidden away within. When a fossil was found we first cut out of the face of the cliff, a place large enough to stand upon, and then carved out the specimen.

I could tell of a hundred narrow escapes from death. One day I was standing on a couple of oblong concretions, about a foot in length, with a chasm, fifty feet deep and three or four feet wide, immediately in front of me. After I had searched carefully the surface of all the rocks in sight, I started to jump over to a narrow ledge on the other side of the gorge. Suddenly both concretions flew from under my feet, and I was plunging head downward into the gorge when by a violent struggle in mid-air I managed to throw my elbows on the ledge; and I hung there until I could find a foothold and pull myself out onto solid rock.

Another time I was climbing a steep slope which was capped by a perpendicular ledge. I thought, however, that I could climb over it to the top of a ridge that ran back into the hills, where

I could find a way down. For understand, we could never go back the way we had come, as we could not relax our muscles sufficiently to enable us to find with the tips of our toes the niches by which we had climbed up. So we had to be sure that we could get to the top and find a way down from there. On this occasion I was so busy searching the face of the rock for fossils that I worked for hours, climbing up niche after niche, without noticing very much where I was going, until chancing to look upward, I discovered that an escarpment of the top ledge leaned over the slope that I was scaling, rendering it impossible for me to reach the top. I fully expected that I should have to cut out a plate to sit in and wait until the boys missed me and looked for me. They could then reach the top of the ledge by some other way, and lower a rope to me. But I was delighted to find at last a perpendicular seam in the rocky ledge, which proved wide enough to admit my body. So I climbed to the top as a man climbs a narrow well, with my back braced against one side and my feet planted against the other.

But such experiences as these, instead of making us timid, only spurred us on to more dangerous attempts. To show how reckless we became, I remember that once Bill found a skull in a perpendicular cliff of solidified volcanic mud, the termination of a ridge that ran far back into the hills. The skull was located about twenty feet up the face of the cliff, and too far below the surface of the ridge to be reached from above; so that there was no way to get at it but by scaling the cliff. I cut niches on one side, and Bill on the other, and we climbed up until we could reach the specimen with our picks, clinging to a niche with one hand and wielding the pick with the other. I worked with my right hand and Bill with his left.

The rock was very hard, and it took a long while to hew out the specimen. While we were at work, we heard a mountain sheep bleating for her young. By reaching up we could get our hands over the edge of the cliff, and pull ourselves up so that we could just peek over. Sure enough, the sheep was coming down the ridge toward us in great excitement, rending the air with calls for her lamb. I began to imitate the bleat of her offspring, and she increased her speed toward us with every sign of relief.

"What if she should butt us off?" I said to Bill, and the position we were in, clinging to the face of the rock with our toes and fingers, made the idea so inexpressibly funny that he began to laugh, louder and louder the more I tried to hush him up. When I had led the sheep up to within ten feet of us, she concluded that we were not her lost lamb, and turning like a flash, started on a run for the mountains a mile away. Out of a side canyon came the lamb and fell in behind its mother; and we could see the dirt flying out behind them until they appeared to be about the size of a rabbit and a ground squirrel.

One day Bill and I were out together in the beds, and when we got back to dinner, Jake did not show up. We were not much concerned about him, as we concluded that he had found a specimen and was digging it out; but when we came in at night and there was still no Jake, we made up our minds that he had either fallen and killed himself or that he was lying in some gulch with a broken limb. In great anxiety we started out into the Bad Lands to find him.

It was a dangerous enough expedition in the daytime, but doubly so at night, and we risked our lives many times; but we did not give up until we had made the desolate region ring with our calls. At last, about midnight, with fear and sorrow in our heart, we returned to camp. By the moonlight I saw what appeared to be a human form in Jake's bed. I rushed to it and threw off the blankets, and there, sleeping peacefully, lay Jake. We had a great mind to take him out into the Bad Lands and pitch him off into a canyon. It seems that he had been to the mountains, three miles away, where a small exposure of the John Day beds could be seen from camp; and when he returned and we were not in, he had not worried about us, but had eaten his supper and gone to bed, while we were making ourselves hoarse shouting for him. This incident illustrates a peculiarity of youth - its thoughtlessness as to the anxiety which it may be causing its elders.

Among the fossil remains which we secured in these John Day beds, were the limbs of a huge *Elotherium humerosum*, so named by Cope on account of the great process on the humerus. We found the specimen in Haystack Valley, lying on its side, with its toes sticking out of the face of a slope. There were thousands of feet of volcanic rock above it. Following in with pick and shovel, we cleaned up the floor, to find, when we reached the center of the humeri and femora, that they had been cut through as smoothly as if it had been done with a diamond saw. I knew, of course, that there had been a fault here, and that the earth in slipping down had severed the bones. The question that interested me was which side had gone down and how far. If the side toward the open valley, then the rest of the skeleton must have been destroyed by the wash, as the slope above the bones lay at an angle of 45 degrees to the floor on which they lay. If, on the other hand, the mountain side had gone down, and the slip had not been too great, I should be able to find the rest of the bones. Inspired by this hope, we put in several days of hard work, and were delighted to find the severed bones three feet below the original level.

What a shaking and trembling of the earth's crust there must have been, when miles of the mountain mass slipped down three feet toward the center of the earth! No wonder that when a similar fault occurred at San Francisco, the puny works of man fell in ruins. The bones of this *Elotherium* are now on exhibition in the American Museum, which purchased the Cope collection, including the material that I secured through eight seasons in the field in charge of his expedition.

I had found in the Cottonwood beds that lie on top of the John Day Miocene the cannon-bone, or long cylindrical foot bone, of a large camel. As I closely studied this bone, which is composed of opposite halves, separated by a thin septum of bone in the center, with a medullary canal on each side, the conviction came to me that the two halves had once been distinct, like the metacarpals and metatarsals of the pig. With this idea in mind, I was constantly looking for a camel in the older beds, and I cannot express my delight when one day, as I was exploring the John Day beds, I came across a skeleton which had been weathered out and lay in bold relief on the face of a slope. I knew before I picked up the cannon-bone that my belief was verified, and when I took up the two bones separately, the fact was proved beyond a doubt that in this ancestor of the living form the metacarpals of the fore foot and the metatarsals of the hind foot were respectively distinct. As the species represented by this specimen was new to science, Professor Cope named it in my honor *Paratylopus sternbergi*. A skull of this species was afterwards found by Dr. Wortman, and both specimens are now on exhibition in the American Museum.

I arrived at this conclusion with regard to the cannon-bone of the ancient camel as Darwin, Marsh, and Huxley arrived at the conclusion that the ancient horse had three toes. They recognized that the split bones of the horse represented the side toes of rhinoceroses, one on each side of the middle metacarpals and metatarsals respectively, and they decided that they were the remnants of side toes in the ancestor of the horse. And later we also found a three-toed horse.

I secured also in these beds the skull of a peccary and an oreodont, both new, and used as the types of Cope's description, and a couple of carnivores; one, called by Cope *Archaelurus debilis*, about the size of the American panther, the other a dog about the size of a coyote. Cope gave the name *Enhydrocyon stenocephalus* to this genus and species. A splendid skull of the rhinoceros *Diceratherium nanum* Marsh, was another of my discoveries here. All the specimens, with the skull of a rodent from the same beds, are now on exhibition in the American Museum.

Of course these are but a few of the many specimens secured in these beds; hundreds are stored away in the drawers and trays of the Museum. I was told that it would cost twenty-five dollars to get a typewritten copy of the list of John Day fossils in the Museum. In that list are many specimens which my party secured or which I purchased from Warfield and Day. Professor Cope once wrote me that my collection there represented about fifty species of extinct mammals.

One day in July I left Jake Wortman in the field and started for Dayville, leading a pack pony. I intended to stay all night with Mr. Mascall, leave my load of fossils, and take back a load of provisions. Bill Day had lost one of the horses, and as a large band of Umatilla Indians was encamped on Fox Prairie at the summit of the mountains, about six miles east of our camp in the Cove, he had gone off in that direction to look for it.

When I reached the high mountain above Dayville, I could look down into the narrow valley of the John Day. Although it was noon, there was no smoke rising from the chimneys of the houses. The fields of wheat were ripe for the cradle - they had no machines in that region, and not only cradled their grain, but threshed it with horses, who tramped it out - but no one was working in them, and there was no stock in the pastures. What could it mean? I asked myself; and as I followed the long trail down to the river, my heart was full of fearful forebodings. Had a pestilence killed all these people whom I knew so well? Or had they all fled, with their horses and cattle, from Indians on the warpath?

Without expecting to hear a response, I called, when I reached the river, for Mr. Mascall to come over with his boat and take me across. To my delight, I saw him come out of his house and take the trail down to the boat through the woods that covered the first river bottom. All the while that he was unlocking the boat and rowing across, I kept shouting, "What's the trouble? Where are all the people?" But not until I had got aboard with my pack and saddle, and we had started back, he answer the questions which I had been asking myself ever since I left the top of the mountain.

It seems that three hundred Bannocks, or Snakes, under their chosen leader, Egan, had left the Malheur Agency, several hundred miles south, and after stealing six thousand horses, mainly from the French brothers' ranch, were now on their way north to join Homely, the chief of the Umatilla, at Fox Prairie. General Howard, who was in hot pursuit, had sent a courier ahead of his command to the settlers in the John Day valley, advising them to gather at some central locality, build a stockade, and take their women and children into it for protection from the treacherous redskins. Everyone in the valley, except Mr. Mascall and an old man who kept the mail station on Cottonwood Creek, a mile to the south, had taken this advice and gone to Spanish Gulch, a mining town on top of the mountains about ten miles southwest.

Near sundown Bill Day came in, having heard the news at the Indian camp. He instantly insisted that we leave everything and go to Spanish Gulch. It was foolish, he said, to risk our lives going back to warn Jake. On the long trail up the mountain we should be in full sight of the South Fork, down which the Indians were expected to come, and it would take us half a day to climb those four thousand feet and hide ourselves in the canyons on the other side. I refused, however, to be moved by his arguments. I told him that I meant to go back, and that he was to go with me. We could not leave Jake there in camp, entirely unconscious of the fate that might be approaching him. He knew nothing of the proximity of hostile Indians, and it was our duty to warn him.

"Well," Bill said, "I am going to look out for number one. I have not lost any Indians. If you have, go and hunt trouble. Let Jake look out for himself."

All my shells, perhaps three hundred, were empty, but I had plenty of powder and lead, and the best long-range rifle I had ever owned, a heavy Sharp's weighing fourteen pounds, and shooting a hundred and twenty grains of lead and seventy grains of powder. I set to work cleaning and oiling it; and then spent the whole night in front of the fireplace, melting lead, casting bullets, and loading shells. Bill also stayed awake, and with his needle-gun kept guard at a porthole which commanded a good view of the open ground around the house.

The next morning I started alone on my pony to follow the trail to the Cove, where Jake, unconscious of danger, was at work in the fossil beds. It seemed an interminable journey, and I thought that there was an ambuscade behind every bush and pile of rocks that guarded the road. But, greatly relieved, I got out of sight at last in the deep canyons on the other side, and soon saw Jake's pony near a fossil bed and found Jake himself deeply interested in a splendid discovery he had made.

When I told him the news, he wanted to drop everything until the war was over, and fly for safety to the stockade. But no; my tent, with many fine fossils in it, was in an open valley in plain sight for miles, and would quickly attract any marauding hostile, who might set fire to it and destroy the work of months. I insisted, therefore, upon caching, the Pacific coast term for hiding, everything. So we took down the tent, and putting it, with the fossils and all the rest of the outfit, into a secret place, we covered them with a big brush pile. Then I was ready to fly as fast as our ponies could carry us.

When we reached the river, Bill was still with Mr. Mascall, and brought over the boat. Then both men insisted that we go without further delay to the Gulch, as we had risked our lives long enough. But there was a large collection of valuable fossils in the log house behind Mr. Mascall's cabin, and as the specimens were wrapped in burlap, they would be destroyed if the Indians burned down the house, which they would be sure to do if they came. I had no boxes, but I had a quantity of new lumber, which Nye had secured from a mill in the vicinity; so, refusing to be moved, I took off my coat and went to work sawing up the lumber and making boxes. The other men never let their guns leave their hands, and kept guard till night, expecting every moment to hear the whoop of the Indians.

By daylight I had every fossil neatly packed, each in a little box, and then we all took hold, and carrying the boxes down to the first river bottom, bid them under a great grapevine, which completely covered them. After throwing dead leaves over our trail, I was satisfied that we had done all that we could, and as we could not induce Mascall to abandon his property, we left him and went over to the Gulch. We found nearly all the settlers keeping house inside the stockade, which was built of pine logs and covered enough ground to hold their teams, wagons, and cattle, as well as themselves.

As I realized that it would be impossible for us to do any work in the John Day beds, fearing every moment to be surprised by Indians, I concluded that this would be a good time to go to the Dalles and try to find out what had become of the collection of Fossil Lake material which had been sent off the year before, and had been lost somewhere. I had a receipt for the specimens from a Mr. French, who was, I supposed, the agent for the Oregon Steam Navigation Company. His letterhead read "Forwarding Agent for the 0. S. N. Co.," but I had repeatedly written to the agent at the Dalles, and had received no answer, while Cope, from his end of the line at Philadelphia, had sent tracers out over every route he could think of, trying to locate the fossils.

A Mr. Wood, the owner of a large herd of horses, was driving the herd to a point near the Dalles for protection from the Indians, and I joined his party. But the several hundred horses raised such a volume of dust that, after a few days of suffocation, I concluded that I might as well lose my scalp as be choked to death, and leaving the herd, went on alone. All along the way, men, women, and children were fleeing for safety to the Dalles, and dozens of homes and ranches were being deserted just at the time when the people should have been saving their grain. I never in my life saw so much excitement and fear. As many white men were fleeing for their lives as there were Indians on the warpath, and every man of them was blaming General Howard for not having exterminated the hostiles before they started.

I met the man who had hauled my Fossil Lake collection in to the Dalles, and for the first time learned the truth about them. It seems that they had never been shipped. Mr. French simply had a warehouse, and forwarded goods by the Steam Navigation Company, and mine had been covered up in the warehouse and entirely forgotten. I was in splendid spirits when I knew that they were safe.

Having rescued this valuable material from the warehouse, I returned to the Gulch without seeing an Indian, to find the people still in a state of great excitement. General Howard had sent word that the men could put themselves under the leadership of Colonel Bernard, each citizen furnishing his own mount and arms, but receiving his rations from the Government. I tried to raise a company of men to accept this offer, but not a man cared to go. At last, heartily tired of staying in camp,

I asked for a volunteer to go with me to the John Day valley to find out how Mr. Mascall and the old man at the stage station were getting on. No one would go at first, but later Mr. Leander Davis, who was for many years a fossil hunter for Professor Marsh, agreed to go with me; and packing a horse with blankets and supplies, we started.

We were relieved to find both men well, and no sign of Indians. Continuing our journey east, we crossed the south fork of the John Day, and all doubts as to the movements of the Indians were removed. For a wide trail, cut deeply into the dry soil by six thousand horses and the three hundred Indians who were driving them north, led down the slope and followed up the main fork on the Canyon City road.

As we sat on our horses, looking south along the heavy trail, we saw some half-dozen horse men coming toward us. We knew that they must have seen us, and concluded to stay where we were until we could make them out. Before long we saw the glitter of sabers and the flash of gold buttons, and soon General Howard and his staff rode up at a gallop. I recognized him by his brigadier general straps and by his empty sleeve. He had lost an arm fighting to preserve the Union.

We saluted, and he asked me whether we had seen his pack train. When I answer no, he asked me if we knew where he could find some bacon, a he and his staff, as well as the troops behind them, had been living for three days on fresh beef without any salt. I told him of a smokehouse across the bridge, and he sent his scout to examine it. The man returned shortly with the report that not only was the smokehouse full of bacon, but that the table in the dwelling house was set for a meal, with cold coffee in the cups, bread, cold bacon, and potatoes, all ready to eat. The people had evidently just sat down to dinner when someone had rushed in with the news that the Indians were coming, and they had all thrown back their chairs and fled for their lives.

While the General and his staff sat down to a hearty meal, Leander and I continued to follow the trail. At one place, where a farmer made cheese, we found that a number of large cheese had been taken out into the road and rolled along for some distance with a stick. We followed up the trail which they had made in the deep dust, and put one of them on our pack. We went into one of the houses on the road, and found that the Indians had broken up all the furniture, including the sewing machine, etc. In the front room they had poured out a barrel of molasses, spread over it several sacks of flour, and stuck a little woolly dog in the mixture. The poor little fellow was dead. A little farther on, a sheepman's house had been burned, and near by two thousand sheep had been mutilated and thrown into piles to die. The herders were found scalped a few days later. At one farmhouse a fine brood mare had been killed because she could not keep up with the herd.

Fig. 29.–**FOSSIL-BEARING CLIFFS.** (After Merriam.) Mascall and Rattlesnake formations.

Fig. 30.–**FOSSIL-BEARING CLIFFS.** (After Merriam.) The Palisades of the Clarno formations.

Some days later, on the twenty-ninth of July, I believe, there was a total eclipse of the sun. The heavens were like brass, and there was a peculiar condition of the atmosphere such as I have never experienced before or since. A report was spread abroad that the Indians had returned and burned all the farmhouses along the river, I was at the time with Leander Davis, and we rode up to Perkins ranch where a lot of men had congregated and were taking turns standing guard for fear of the Indians. When we rode up they were standing about, uncertain as to what it all meant. The dogs had gone under the stoop and the chickens to roost. The air was motionless, and an unusual stillness was over everything. The men welcomed us in hushed voices.

I sprang from my horse and asked Perkins whether he had any pieces of broken glass. He said that there were plenty under the west window, and I went and got a supply, followed by all the men, who were greatly relieved by my explanation of the phenomenon. We got a candle and blackened the pieces of glass, and watched the progress of the eclipse through them.

It had a more disquieting effect upon the hostile Indians. It seems that the soldiers had cut them off from crossing the Columbia by capturing all the small boats and patrolling the river night and day; so that with Howard's troops on the trail behind them, troops from Walla Walla on their flanks, and the river in front, they were in a bad way. Moreover, the French brothers and the governor of Oregon had offered a reward of two thousand dollars for Egan's head.

The Umatilla Indians were accused of pretending to help the whites in the daytime, and really helping the Snakes at night. So the commander sent out a party of soldiers to capture the squaws and little children of Homely and the other chiefs and hold them as hostages for the good behavior of their braves. When the latter asked the commander to release their families, the answer was given that if they would capture Egan and deliver him up to the authorities, they would not only get back their wives and children, but would receive the two-thousand dollar reward. Otherwise their families would still be held as hostages.

It appeared that Egan had an appointment with Homely at a certain hour. As he rode out from his camp, with a brave behind him, Homely, similarly attended, went out to meet him. When they met between the two camps, they turned at right angles and rode toward the point agreed upon for the powwow. But as they were riding thus, side by side, Homely, with a word to his brave, suddenly raised his rifle and shot Egan, while his brave shot the attending Snake. They then immediately severed the heads of the dead men, and riding back with them to the whites, claimed the reward. About the same time, the eclipse came on. And the poor Snakes, deprived of their leader, thought that the world was coming to an end, and leaving their great herd of stolen horses, fled in small bands toward the Malheur Reservation and were all eventually captured.

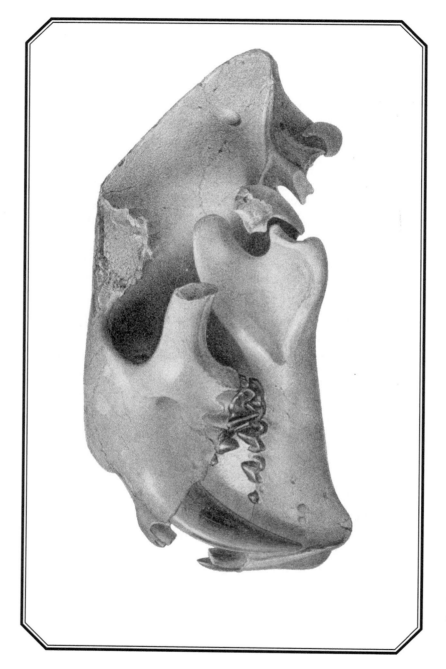

*Fig. 31.—*Skull of Great Sabre-toothed Tiger, *Pogonodon platycopis.*
Discovered in John Day River, 1879, by Leander Davis. (After Cope.)

The war thus ended, as soon as I could get things in shape and my party together, I returned to the Cove, got my outfit and fossils, and moved over into Haystack Valley. I remained there all winter, and the next season secured another large collection. Many of the specimens in it are described by Professor Cope in Vol. III of the "Tertiary Vertebrata." On p. xxvi and the two following pages of the preface, he pays his collectors a high compliment, which I give myself the pleasure of repeating here in his own words: "The same year ['77] I employed Charles H. Sternberg to conduct an exploration of the Cretaceous and Tertiary formations of Kansas. After a successful search, I sent Mr. Sternberg to Oregon. The John Day formation was chiefly examined on the John Day River and the Loup Fork beds at various points in the same region. These yielded about fifty species, many of them represented in an admirable state of preservation."

After mentioning the work of his other explorers, he goes on to say: "Mr. Sternberg's expedition of 1878 was interrupted by the Bannock war and both himself and Mr. Wortman were compelled to leave their camp and outfit in the field and fly to a place of safety on their horses. It is evident that an enthusiastic devotion to science has actuated these explorers of our western wilderness, financial considerations having been but a secondary inducement. And I wish to remark that the courage and disregard of physical comfort displayed by the gentlemen above referred to are qualities of which their country may be proud, and are worthy of the highest commendation and of imitation in every field."

Before leaving this interesting field, I wish to show my readers Cope's figure of the great saber-toothed tiger, *Pogonodon platycopis* (Fig. 31), which was secured in 1879 by Leander Davis. I do not remember who first discovered the specimen, but for weeks each of us collectors, Wortman, Davis, and I, tried to devise some means of securing it. The skull topped a pinnacle, perhaps thirty or forty feet high, and tapering like the spire of a church. At the top it was only a foot in diameter. We knew that it would not be strong enough to support the weight of a ladder, and it was too steep to scale. Moreover, if we blew it up with powder, the skull, whose rows of teeth seemed to grin at us defiantly, would be shattered to bits.

By whatever method it was secured, it represented a feat of the greatest possible bravery, and Cope did only justice to Leander Davis in publishing his understanding of the manner in which it was done. That description is attached to the skull to-day, and thousands have read of Davis' heroic act in securing it for science. Professor Cope says that he cut niches and climbed to the top of the spire. My remembrance, however, is that he threw a rope around the spire and let it settle down to where he thought the rock would be strong enough to support his weight. He then climbed up hand over hand to the loop, stood erect, picked up the skull, and without putting any pressure on the rock, got back to his rope and down to safety below. He then secured the rope by jerking off the top of the pinnacle.

It matters little how he got the skull, but I am ready to testify that it was the bravest undertaking I ever saw accomplished in the John Day beds; and as long as science lasts, this noble specimen of one of the largest tigers that ever lived should be associated with the name of Leander Davis. I am glad that the great dike across the Cove is named after him also.

What is it that urges a man to risk his life in these precipitous fossil beds? I can answer only for myself, but with me there were two motives, the desire to add to human knowledge, which has been the great motive of my life, and the hunting instinct, which is deeply planted in my heart. Not the desire to destroy life, but to see it. The man whose love for wild animals is most deeply developed is not he who ruthlessly takes their lives, but he who follows them with the camera, studies them with loving sympathy, and pictures them in their various haunts. It is thus that I love creatures of other ages, and that I want to become acquainted with them in their natural environments. They are never dead to me; my imagination breathes life into "the valley of dry bones," and not only do the living forms of the animals stand before me, but the countries which they inhabited rise for me through the mists of the ages.

The mind fills with awe as it journeys back to those far-distant lands. Stop, reader, and think! In this John Day region, ten thousand feet, or nearly two miles, of sedimentary and volcanic rock lie above the Niobrara Group of the Cretaceous, from which I dug last summer the beautiful skull of a Kansas mosasaur, *Platecarpus coryphaeus*, which lies before me now, its glistening teeth as perfect as in the days when they dripped with the blood of its victims. How many ages were those ten thousand feet in building? How long has it taken the running water, with its tools of sand and gravel, to carve out the Grande Coulée and the river valley, and expose all the various formations, with their records of the life of the past? And yet all this has taken place since my mosasaur, which seems to watch me as I write, fought its last battle and sank to rest beneath the waves of the Cretaceous sea.

A Sternberg

Glossary

◄ A ►

actuate: to move into action

ambuscade: to place an ambush or a surprise attack

amphitheater: a flat or gently sloping area surrounded by abrupt slopes

◄ B ►

bad lands: a region marked by intricate erosional sculpting, scanty vegetation and fantastically formed hills (characteristic of the John Day formations of the John Day Fossil Beds)

basalt: a dark gray to black dense, fine-grained rock formed from cooling lava

brave: an American Indian warrior

beds: layers of rock

◄ C ►

cannon bone: a bone in hoofed mammals that supports the leg from the hock joint to the fetlock

carnivore: a flesh eating animal

chalk: a soft white, grey, or buff limestone composed chiefly of single-celled marine organisms

chasm: a deep cleft in the earth

concretion: a mass of mineral matter found generally in rock of a composition different from its own and produced by deposition from groundwater in the rock

cordwood: cut wood stored in a pile and intended for fuel

coulée: the French word for ravine. In the past, Picture Gorge was referred to as the Grande Coulée.

courier: a member of the armed services whose duties include carrying mail, information or supplies

cradle: a frame of wood used in agriculture for cutting and storing grain

Cretaceous: the last period of the Mesozoic era, from 144 to 65 million years ago; the last period in which dinosaurs existed

      The Life of a Fossil Hunter

◁ D ▷

dike: a body of rock that cuts across other preexisting layers; commonly igneous rock that has been injected into a fissure

◁ E ▷

escarpment: a long cliff or steep slope separating two comparatively level or more gently sloping surfaces and resulting from erosion or faulting

excavate: to dig out and remove, specifically referred to here as the removal of fossils

◁ F ▷

fathoms: a unit of length equal to six feet used especially for measuring the depth of water

fault: a fracture in the earth's crust accompanied by a displacement of one side of the fracture with respect to the other and in a direction parallel to the fracture

femur: the proximal bone of the hind or lower limb, thighbone

fetlock: a projection bearing a tuft of hair on the back of the leg above the hoof of a horse or a similar animal

flora: a compilation of the plants of an area or period

formation: a group of related rock layers extensive enough to be mapped

fossil: any evidence of past life preserved by geologic processes

◁ G ▷

genus: a class, kind, or group marked by common characteristics

group: a series of related rock formations

gulch: a deep or precipitous cleft

◄ H ►

herbivore: a plant eating animal

hew: to cut with blows of a heavy cutting instrument

humerus: the long bone of the upper arm or forelimb extending from the shoulder to the elbow

◄ I ►

interminable: having or seeming to have no end

◄ J ►

jints: slang term referring to bits and pieces of joints

◄ K ►

knucks: slang term referring to bits and pieces of joints, such as knuckles

◄ L ►

larder: pantry of food

limestone: a sedimentary rock consisting chiefly of calcium carbonate often derived from the accumulation of shells or coral

locality: a particular place, situation or location; in paleontology, a place where fossils have been found

◄ M ►

mammoth: a cold adapted elephant that lived during the Pleistocene Epoch from 2 million to 10 thousand years ago

manifested: to make evident or certain by showing or displaying

marauding: to roam about and raid in search of material goods to steal or take advantage of

medulary: the most inner part of an animal bone

metacarpals: the bones in a the hand or fore foot between the wrist and digits

metatarsals: the bones in a foot or hind foot between the ankle and digits

Miocene: an epoch of the Tertiary Period between the Pliocene and the Oligocene or the corresponding system of rocks, dating from 23.8 to 5.3 million years ago

molasses: a syrup made from boiling down sweet vegetable or fruit juice

molten: liquefied by heat

mosasaur: a marine reptile that lived in near shore ocean waters during the Late Cretaceous Period

native: living or growing naturally in a particular region

niche: a space suitable for that which fills it

oblong: rectangular and longer in one direction than another

Oreodon: (also oreodont) a type of hoofed North American mammal resembling both a pig and a sheep that became extinct 12 million years ago

organic: derived from a living or once living organism

panorama: an unobstructed or complete view of a region in every direction

peccary: a North American mammal resembling a pig

pestilence: a contagious or infectious disease that is virulent and devastating

phalanges: the digital bones of the hand

placer: a surface mineral deposit formed from the concentration of mineral particles by mechanical weathering. A common example of a placer deposit is gold nuggets which become deposited in stream gravels.

precipitous: having a very steep ascent

process: a prominent or projecting part of an organism or organic structure (bone)

protozoan: a single-celled animal

proximity: very near or soon forthcoming

◄ Q ►

quadruped: having four feet

◄ S ►

saber: a cavalry sword with a curved blade

sandstone: a sedimentary rock consisting of cemented sand sized rock particles which are usually particles of quartz

sediment: rock or mineral particles that are transported by water or wind

sedimentary: pertaining to or containing sediment, or formed by its deposition

septum: a dividing wall or membrane between two bodily spaces or masses of soft tissue

shod: furnished or equipped with a shoe

species: a category of biological classification indicating populations potentially capable of interbreeding

splint: either of two small, thin bones on the back of each foot bone on a horse or related animal

stockade: a line of stout posts set firmly to form a defense

squaw: a disrespectful term used extensively in the past, referring to an American Indian woman

stoop: raised platform, stairs, or porch attached to building

◁ T ▷

tarsals: see tarsus

tarsus: bones relating to the flat of the foot and ankle

Tertiary: the geologic period following the extinction of the dinosaurs, dating from 65 to 0.01 million years ago which is often referred to as the "Age of Mammals." The Tertiary is divided into five epochs: Paleocene, Eocene, Oligocene, Miocene, Pliocene and Pleistocene.

tracers: requests for information by message

◁ V ▷

venison: the flesh of a deer

vertebrae: one of the bony segments composing the spinal column

vertebrate: any animal having a back bone, including fish, amphibians, reptiles, mammals and birds

vulcanism: processes by which molten rock within the earth rises and is extruded onto the surface, also spelled: volcanism

◁ W ▷

wash: loose or eroded surface rock transported and deposited by running water

weathered: alterations in color, texture, composition or form by exposure to air, water, and surface processes

RESOURCES FOR FURTHER STUDY

◄ BOOKS ►

The Best Book of Fossils Rocks, and Minerals
by Chris Pellant

The Big Cats and Their Fossil Relatives
by Alan Turner, F. Clark Howell,
illustrated by Mauricio Anton

Common Fossil Plants of Western North America
by William D. Tidwell

Dinosaur Dynasty
by Katherine Rogers

Gilded Dinosaur:
The Fossil War Between E. D. Cope and O. C.
Marsh and the Rise of American Science
by Mark Jaffe

John Day Basin Paleontology
Field Trip Guide and Road Log
by Theodore Fremd, Ericks A. Bestland,
and Gregory J. Retallack

Life of a Fossil Hunter
by Charles Sternberg

National Audubon Society Field Guide to Fossils
by Ida Thompson

The Oligocene Bridge Creek Flora of
the John Day Formation, Oregon
by Herbert W. Meyer and Steven R. Manchester

Strange Genius:
The Life of Ferdinand Vandeveer Hayden
by Mike Foster

◄ WEBSITES ►

John Day Fossil Beds National Monument
www.nps.gov/joda/

University of California at Berkley
Museum of Paleontology
www.ucmp.berkeley.edu

American Museum of Natural History
www.amnh.org

Sternberg Museum of Natural History
www.sternberg.fhsu.edu

Yale University,
Peabody Museum of Natural History
www.peabody.yale.edu